Yellow Umbrella Books are published by Capstone Press
151 Good Counsel Drive, P.O. Box 669, Mankato, Minnesota 56002
www.capstonepress.com

Library of Congress Cataloging-in-Publication Data
Jacobs, Daniel (Daniel Martin)
    Count your chickens / by Daniel Jacobs.
    p. cm.
    Summary: Simple text and photographs introduce the concept of "skip counting," or counting by groups of a certain number, and present items for practice.
    ISBN 0-7368-2918-0 (hardcover)—ISBN 0-7368-2877-X (softcover)
    1. Counting-Juvenile literature. [1. Counting.] I. Title.
QA113.J36 2004
513.2'11—dc22   2003012335

Editorial Credits
Editorial Director: Mary Lindeen
Editor: Jennifer VanVoorst
Photo Researcher: Wanda Winch
Developer: Raindrop Publishing

Photo Credits
Cover: DigitalVision; Title Page: Jerry Tobias/Corbis; Page 2: Kent Knudson/PhotoLink/Photodisc; Page 3: Corel; Page 4: AG/elektravision; Page 5: AG/elektravision; Page 6: Barbara Penoyar/Photodisc; Page 7: Barbara Penoyar/Photodisc; Page 8: PhotoLink/Photodisc; Page 9: PhotoLink/Photodisc; Page 10: Ron Chapple/Thinkstock; Page 11: Ron Chapple/Thinkstock; Page 12: H. Prinz/Corbis; Page 13: H. Prinz/Corbis; Page 14: Stockbyte; Page 15: Stockbyte; Page 16: Mark A. Johnson/Corbis

1 2 3 4 5 6 09 08 07 06 05 04

Count Your Chickens

by Daniel Jacobs

Consultants: David Olson, Director of Undergraduate Studies, and Tamara Olson, PhD, Associate Professor, Department of Mathematical Sciences, Michigan Technological University

Yellow Umbrella Books

an imprint of Capstone Press
Mankato, Minnesota

Count your chickens one by one.

1, 2, 3 is how it's done.

You can also count things two by two.

2, 4, 6 is what you do.

Here are twins. They make a pair.

Count by twos. How many
are there?

You can count these chicks at five in a coop.

How many chicks are in the whole group?

This sea star has five arms, too.

5, 10, 15—you know what to do!

You can count these toes in groups of ten.

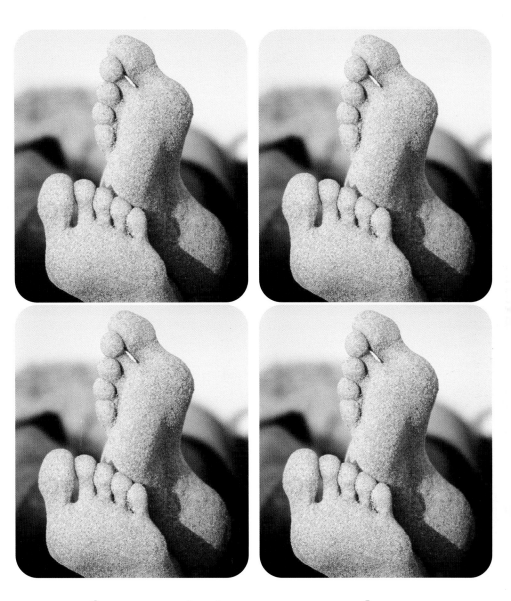

If you didn't count forty,
try again!

Ten geese are flying
through the air.

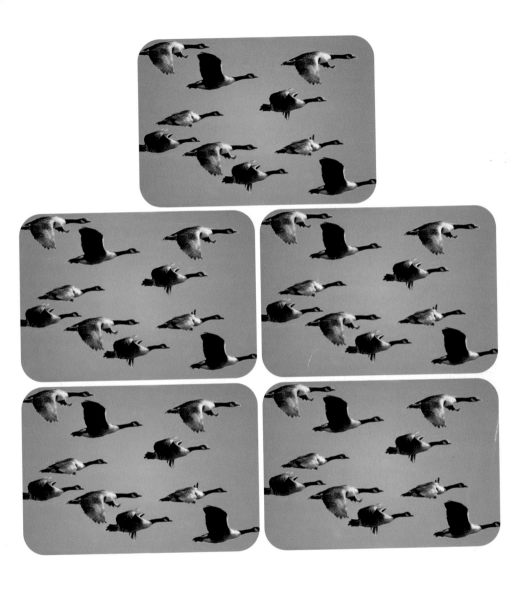

Count by tens. How many are there?

Look at all these fingers in the sun. You can count them many ways. It's so much fun!

Words to Know/Index

Word Count: 123
Early-Intervention Level: 9